Dancing Curves

A Dynamic Demonstration of Geometric Principles

Merwin J. Lyng

Mayville State College
Mayville, North Dakota

National Council of Teachers of Mathematics

Library of Congress Cataloging in Publication Data:

Lyng, Merwin J.
Dancing curves.
1. Conic sections – Models. I. Title.
QA559.L96 516'.23 78-2781
ISBN 0-87353-124-8

Printed in the United States of America

Table of Contents

Preface

During the past few years an illustrated lecture, "Dancing Curves," has been presented at national, regional, state, and local meetings of teachers of mathematics. "Dancing Curves" has also been presented to college and high school student groups and to fraternal and social organizations. It has proved to be an exciting visual experience for nonmathematicians as well as for students of mathematics, science, and art.

Originally conceived as a means of illustrating conic sections—ellipses, parabolas, and hyperbolas—"Dancing Curves" has evolved into a dynamic way of illustrating many lines, curves, and surfaces. Favorable public reaction to this lecture-demonstration has prompted me to write this fairly detailed account of the construction of the apparatus or string model used in the demonstration, together with a suggested procedure for using it.

I am indebted to Hermann Baravalle for the original inspiration that led to the development of "Dancing Curves." I am also grateful to the many students, teachers, and colleagues who have made recommendations for the improvement of the demonstration. Their suggestions have led to some of the variations that are discussed. I am indebted, too, to my family for their encouragement and enthusiastic response as "Dancing Curves" envolved from a crude idea to an exciting visual experience.

Shopping List

Quantity	*Description*	*Price*
1	3/8″ plywood (approximately 2′ × 4′)	$3.20
1	3/8″ galvanized pipe, 105 cm long	2.00
1	3/8″ pipe coupling	0.50
2	3/8″ floor flanges	4.50
3	springs, 10 cm long	1.65
3	#14 screw eyes	0.50
1	ball of #10 parcel post twine	0.75
1	set screw	0.20
	TOTAL	$13.30

Introduction

It has long been known that certain geometric concepts and designs can be illustrated with string models. Many of these models emphasize artistic design as well as the mathematics involved. String models help illustrate the three-dimensional nature of many curves. Surfaces commonly called "ruled surfaces" can be readily illustrated with string models, since the strings tend to represent straight lines.

When light beams are projected through models made of reflective strings, some of the light is reflected by the strings and an image in space is created. The nature of the image depends on the arrangement of the strings and the nature of the light beam. For example, when a light beam representing a plane is passed at right angles through a string cylinder, the image created is a circle.

In the pages that follow, directions will be presented for building a string model that can be used to illustrate curves, surfaces, and solids.

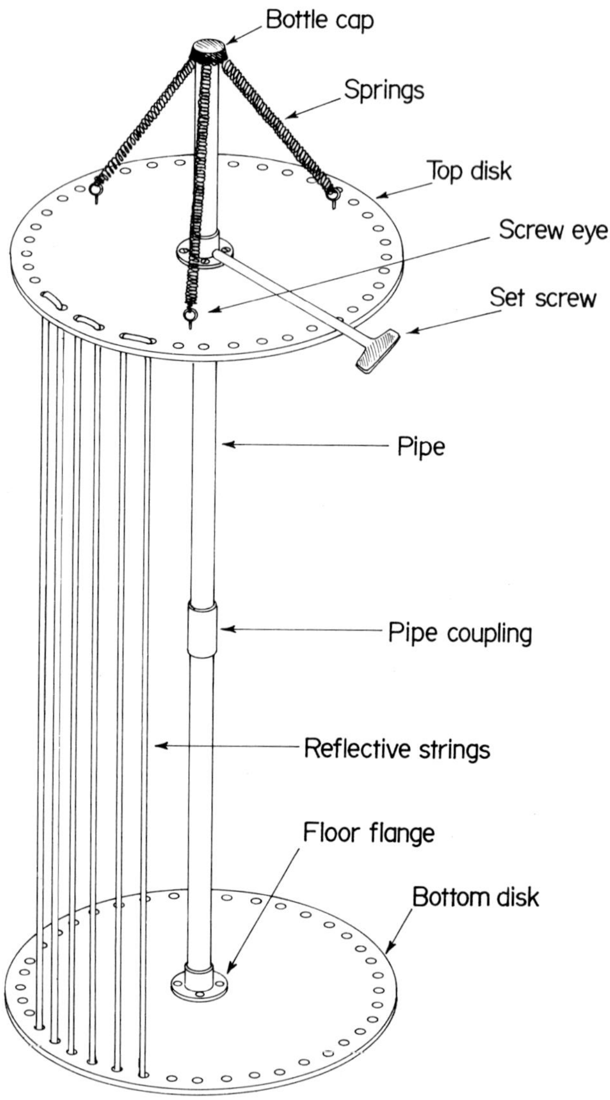
Bottle cap
Springs
Top disk
Screw eye
Set screw
Pipe
Pipe coupling
Reflective strings
Floor flange
Bottom disk

Fig. 1

Making the String Model

The string model described here (illustrated in fig. 1) is 80 centimeters tall and 46 centimeters in diameter. The bases (top and bottom disks) are made of plywood approximately 1 centimeter thick. The strings, any white or reflective material, are approximately 2 centimeters apart. The center post is a 105-centimeter piece of galvanized pipe, threaded at one end. For ease in disassembling and transporting the model, the center post could be two pieces, each approximately 52 centimeters long, held together with a pipe coupling. The size of the pipe is not critical, but pipe with an outside diameter of about 2 centimeters (called 3/8-inch pipe by plumbers) is very convenient.

A floor flange, available at hardware stores or plumbing shops, should be attached to the center of the bottom disk with wood screws or small bolts (bolts will give more stability and strength).

The disks have holes three millimeters in diameter (or any convenient size to accommodate the string used). These holes, approximately one centimeter from the outer edge of the disks, are placed at intervals of two centimeters. Rather than drilling holes for the strings, one could saw slots in the disks at two-centimeter intervals. Two disks, one with drilled holes and one with sawed slots, are pictured in figure 2.

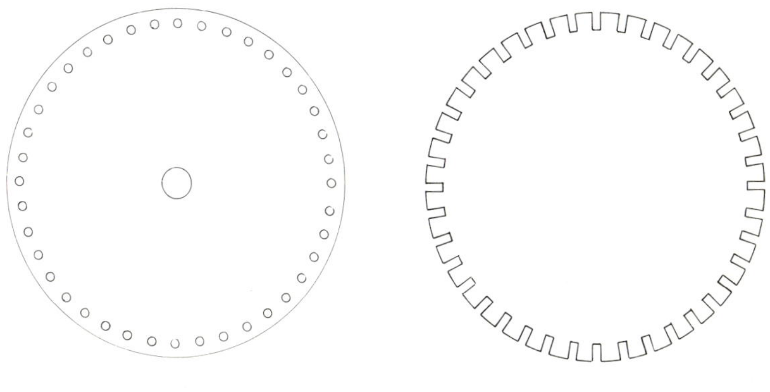

Fig. 2. Disks with drilled holes and sawed slots

The advantage of the sawed slots over the drilled holes is that they make it easier to thread the string between the disks. The disadvantage is that the string may not stay in place without some device to close the slots. One technique might be to put a drop of glue at the entrance to each slot after the string has been completely strung between the disks and the proper tension has been established. Figure 3 illustrates the bottom disk with the floor flange attached and the holes drilled to accommodate the strings.

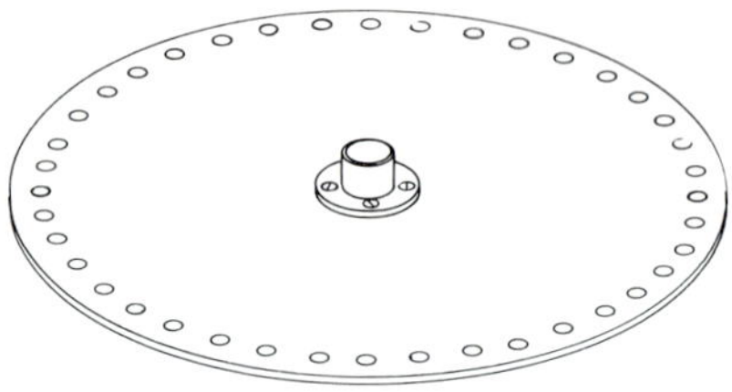

Fig. 3

The top disk is like the bottom one except that it must have a hole in the center to accommodate the pipe. Another floor flange is placed over this hole. This flange must be reamed out so that the pipe can slide through it rather than screw into it. This arrangement allows the pipe to fit snugly inside the floor flange but still allows the top disk to move vertically.

Through the side of this flange is drilled a six-millimeter hole, tapped, or threaded, to accommodate a six-millimeter set screw. Some suppliers may have flanges that do not need to be reamed and that already have holes for set screws. The length of the set screw is not critical except that it should be long enough for easy adjustment. The set screw allows the top disk to be rotated and locked in different positions.

In order to keep tension on the strings, yet allow the top disk to rotate and also to move vertically, springs are attached to the top disk and center pipe. Three number 14 screw eyes are attached to the top disk. They are equally spaced approximately ten centimeters from the pipe. Three springs, each about ten centimeters long and available at most hardware stores, are attached to these three screw eyes. These springs must provide enough tension to keep the strings taut but must also be capable of stretching to about thirty-five centimeters without too much effort. The other ends of the springs are hooked through three small holes, which have been punched or drilled around the edge of a bottle cap.

Select a bottle cap that fits securely over the top of the center pipe yet can turn on the pipe as the top disk is rotated. The cap from a two-liter bleach bottle works fine. Figure 4 illustrates the completed top disk with its floor flange, set screw, and springs in position.

The disks and the pipe can now be painted. If they are painted flat black, they will be nearly invisible in a dark room, and the images created by the light beams will be on the strings alone. However, if the disks and pipe are painted white, they too will reflect the light beams and will illustrate some additional mathematical concepts. One of these concepts will be discussed later.

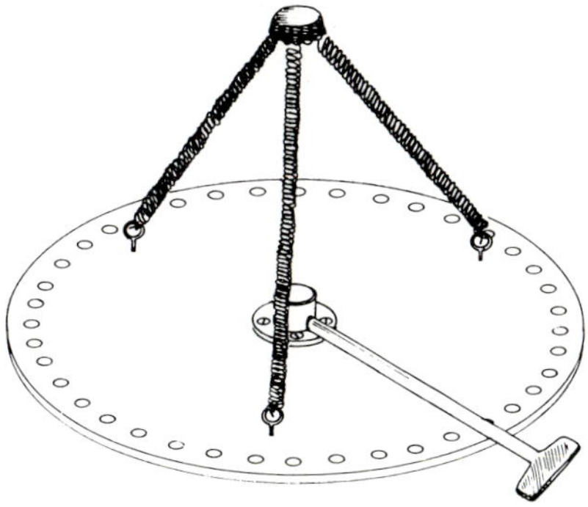

Fig. 4

Once the disks are complete and the spring mechanism and center pipe are ready, you can begin to assemble these parts and prepare to attach the reflective strings.

Begin the assembly by screwing the pipe into the floor flange in the base. Next, place the top disk over the center pipe and fit the bottle cap over the top of the pipe. Push the top disk down approximately twenty-five centimeters below the top of the pipe. Lock the disk in place with the set screw. The disks should now be firmly fixed in place approximately eighty centimeters apart. Figure 5 illustrates the device at this stage of development.

A note of caution might be in order at this point. Although the pipes need not be tightened with a pipe wrench, they must be tight enough to prevent their becoming unscrewed when the model is being used. The gradual loosening of the center pipe can cause the device to malfunction and embarrass the user.

Finally, the string can be attached. This can be done by weaving it up and down through the holes or slots in the two disks. Ten-ply parcel post twine is a good reflective string, although other types may be satisfactory. Certain types of monofilament fishline have been used with success, and fluorescent string can also be used.

The kind of knot used to tie the ends of the string and the position of the knot are not important, since the knot will rarely be visible when the device is used in a dark room. However, it is desirable to have the knots

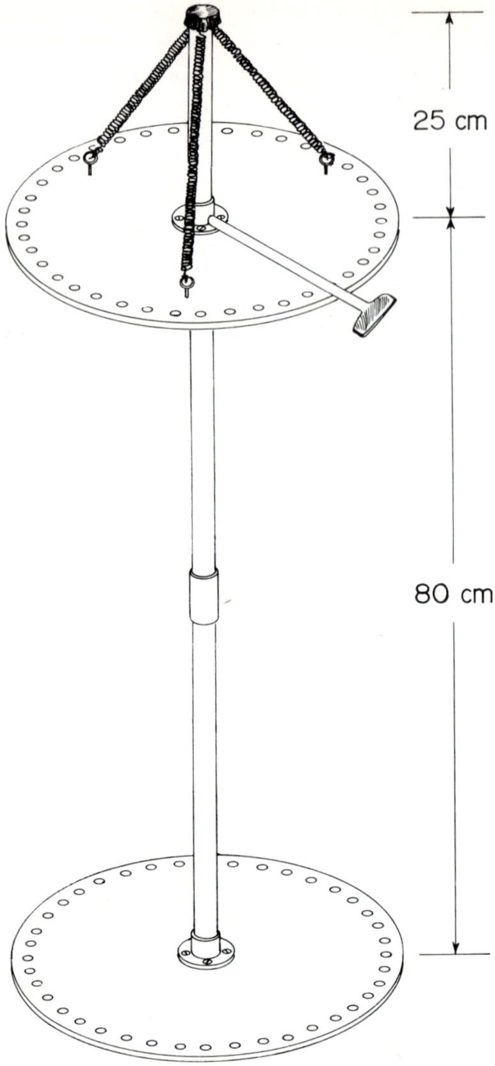

Fig. 5

near the disks. Care should be used to ensure that the knots do not become untied while the device is being used. A drop of glue on the finished knot might be good insurance.

The important points to remember in selecting the string are that it must reflect light and must have a cross section large enough to create an image visible from about ten meters when light strikes it. About fifty-five meters of string is needed for a model of the size described here.

The string model is now ready to use. In its present form it represents a cylinder. When the top disk is rotated about 90 degrees and locked in position, the cylinder is transformed into a hyperboloid of one sheet. When it is rotated 180 degrees and locked, the two nappes of a cone are formed. The cylinder, hyperboloid, and cone are illustrated in figure 6.

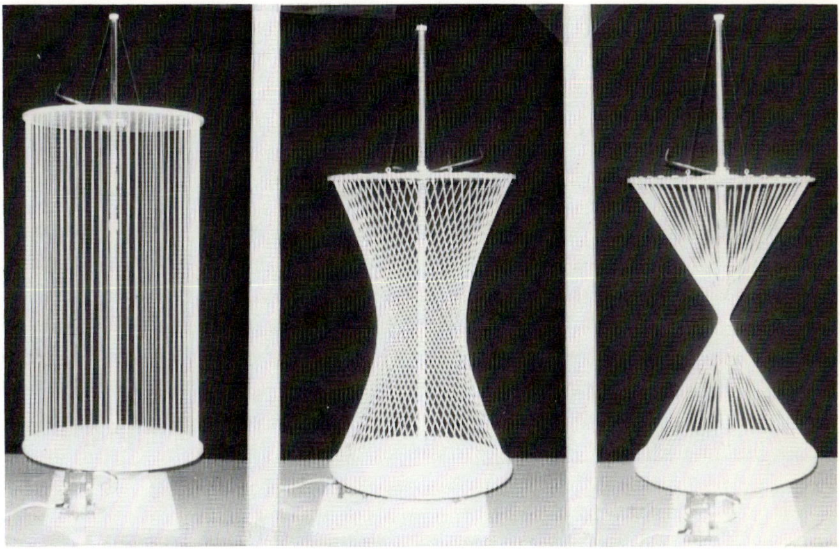

Fig. 6. Cylinder, hyperboloid, and nappes of a cone

This model is an appropriate size for classroom or large-group demonstrations. A smaller one might be more appropriate for small groups.

Making the Slides

Now that the string model is complete, it remains to devise a technique for passing beams of light through the strings to indicate (create) various sections.

A 35-mm slide projector with appropriate slides is quite satisfactory for projecting beams of light. A lantern slide projector will also work but is more cumbersome and usually less readily available. An overhead projector can also be used. This can be done by cutting a line or a curve in a piece of paper so that light passes through the line or curve when the paper is placed on the overhead projector.

A variety of techniques can be used for producing light beams. Let us examine one simple and inexpensive method, which makes use of underexposed photographic slides.

If a flashbulb fails when you are taking a slide picture, the slide is totally black when it comes back from the processor. Such a slide has been underexposed. If you look at it carefully, you will see that one side is dull and the other is shiny. The dull side, which has the photographic emulsion on it, is called the emulsion side. The shiny side is simply the plastic carrier that holds the emulsion.

By taking a sharp instrument, such as the point of a compass, you can make scratches in the emulsion. The scratches must be wide enough to allow an adequate beam of light to pass through the slide when it is placed in the projector but narrow enough to create sharp images. A bit of practice and experimenting may be necessary before this technique is perfected.

An abundant supply of underexposed slides is usually available. Ask your students for any they might have, and within a day or two you will probably have more than you can use. For your convenience, four slides have been prepared for you and are placed inside the back cover of this book.

Using the Slides

When a prepared slide is placed in a projector, light will pass through the scratches. If a horizontal line has been scratched on the slide, a horizontal plane of light will be projected. When this horizontal plane of light is caused to intersect the strings of the cylindrical model, a circular cross section will appear. Figure 7 illustrates a slide with a horizontal line scratched on it, and figure 8 illustrates the circular image that is created when light passes through this slide and intersects the strings of the cylindrical model.

Fig. 7

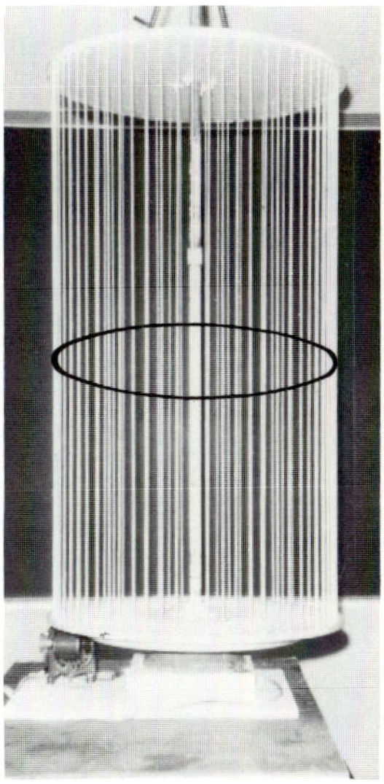

Fig. 8

An elliptical cross section can be produced if the line on the slide is not horizontal. For example, the line on the slide in figure 9 will produce an elliptical image on the string cylinder (fig. 10). Remember that the design on the slide becomes inverted when projected on the strings.

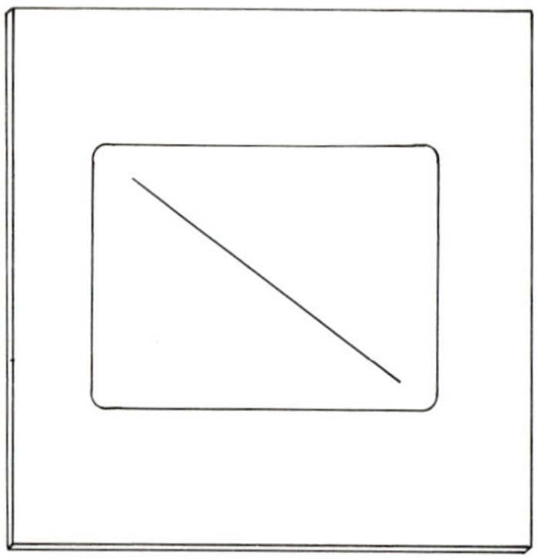

Fig. 9

Fig. 10

By scratching different lines or curves on the slides, you can vary the images projected on the string model. Some of the variations that can be created are illustrated in figure 11. Let your imagination create others.

You need not confine your thinking to the usual geometric figures. Numerals, letters, or even words scratched on the slides make exciting viewing for elementary school children. In fact, any slide picture placed in the projector will create an image on the strings.

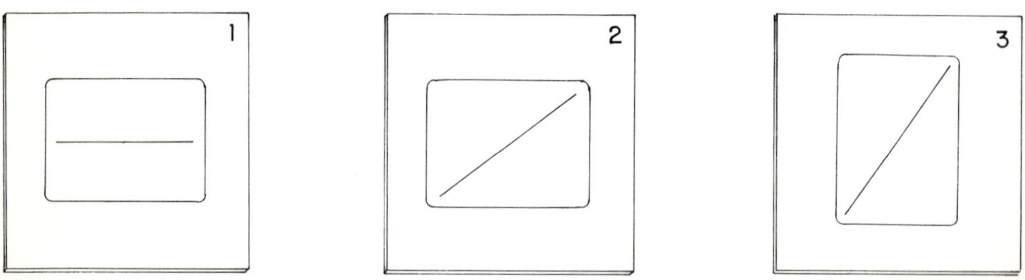

Fig. 11

Some Added Effects

One added effect can be created by coloring some of the slides. The simplest way to do this is to tint the scratch with a felt-tipped pen. One weakness of this technique is that the color fades in time. A more permanent method is to attach color transparency material to the slide after the lines and curves have been scratched on.

Reverse photography is a process by which dark lines on a light background appear as light lines on a dark background. This process can also be used to produce slides. The desired line or curve is drawn on white paper, then photographed.

A photographic slide copier can be used to reproduce the slides created by any of these techniques.

If we change the string cylinder into a hyperboloid of one sheet or into the two nappes of a cone by rotating the top disk, then different images will be created with the sixteen slides illustrated in figure 11. Specifically, when slides 1, 2, 3, and 4 are projected onto the cone, the images will be a circle, an ellipse, a parabola, and a hyperbola, respectively. Each of these conic sections can be dynamically illustrated with this device.

If the center pipe has been painted white, as mentioned earlier, it too will reflect light and become part of the image. With slide 1 the image on the pipe represents the center of the circle. When slides 2, 3, and 4 are used, is the image on the pipe a focal point of the ellipse, parabola, or hyperbola? This question is posed as a challenge to the observer.

If the slide projector is placed on a lazy Susan or some similar device, the light beams can be readily moved from side to side. This lateral motion of the light beams through the strings has the effect of making the images "grow" before the eyes of the observer, and some of the curves appear to "dance" on the strings. Partial cross sections can also be created quite easily in this manner.

An added dimension can be created by placing the string model on a mechanically rotating base. As the light beams intersect the rotating string model, the images appear to be in motion. These moving images can illustrate hyperbolic, parabolic, and elliptical paths or orbits as well as many other moving or dynamic designs. If the base rotates more than three to four revolutions a minute, however, the images will appear blurred.

A lazy Susan turned by a friction drive makes a good base. A friction drive can be made by placing a small rubber wheel on a motor shaft, then placing the motor and rubber wheel so that the friction between the wheel and the edge of the lazy Susan causes the lazy Susan to rotate (fig. 12).

Fig. 12

The number of revolutions a minute that the rubber wheel makes, together with the relative diameters of the lazy Susan and the rubber wheel, will determine the number of revolutions a minute that the lazy Susan will complete. A bit of experimenting may be necessary in order to get the model to revolve at the appropriate speed.

Motors of different sizes and of different power ratings are available at most hobby shops. Many discarded toys have motors that can be used. The motor that drives the spit on a barbeque grill will also work.

The string model can be used to represent a cylinder, a hyperboloid of one sheet, or the nappes of a cone. Hence, there are three different surfaces. Each of these surfaces can either be left stationary or be made to rotate. Thus, there are six situations that can be illustrated with each slide that is projected through the strings: a stationary cylinder, a rotating cylinder, a stationary hyperboloid, a rotating hyperboloid, a stationary cone, and a rotating cone. Multiply these six situations by the number of slides you have made and you will discover the number of different images you can create.

The position of the observer is also a factor, since different images will be observed from different positions with respect to the string model. Conic sections are best illustrated by having the observers, string model, and projector located as suggested in figure 13. Other cross sections can be observed with this arrangement also.

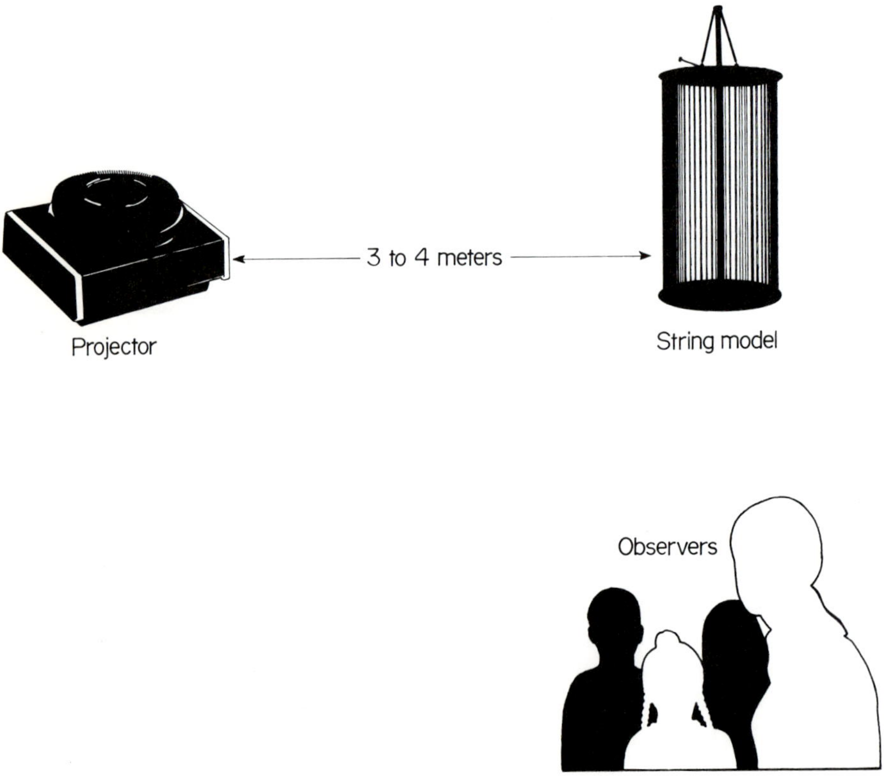

Fig. 13

Additional Uses of the Device

As we have already observed, the string model can be used to illustrate conic and many other sections. Further uses are limited only by the imagination of the user. For example, if the device had been made with triangular bases rather than circular bases, what kinds of cross sections would have been created? What would happen if the string model had square bases? Rectangular bases?

By fixing the circular bases and by placing three sets of strings between one set of disks, we can represent a cylinder, a hyperboloid, and a cone simultaneously. The light beams will illustrate sections of all three string models simultaneously, albeit in a rather crowded fashion.

Another possible variation is to have separate models of the cylinder, hyperboloid, and cone. If the models, projector, and observers are located as suggested in figure 14, the different sections can be seen as the light beam moves from right to left through the strings of each model.

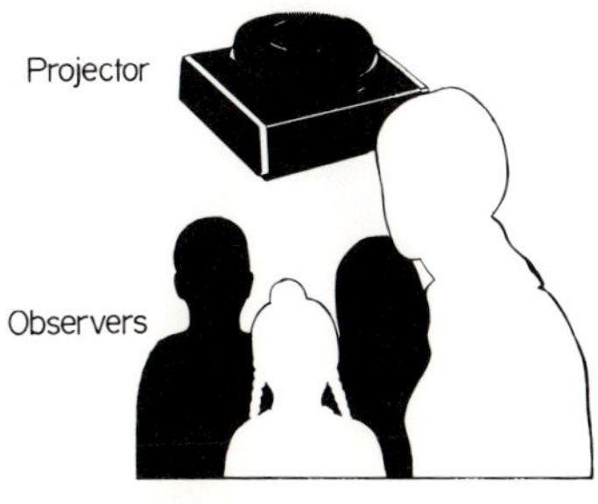

Fig. 14

Conclusion

Although this device was originally conceived as a way of illustrating conic sections, it has proved to be useful for illustrating many other ideas. Experience has shown it to be of interest to the nonmathematician as well as to those who are mathematically inclined. It interests all age groups. It is a source of excitement for those who are interested in artistic design. The circular, elliptical, parabolic, and hyperbolic paths that can be illustrated are useful for depicting many concepts from astronomy. The physics student will observe examples of simple harmonic motion and other concepts from physics. From an artistic point of view, it is an exciting device for illustrating three-dimensional design: in rotation, the device creates three-dimensional design in motion. And, of course, there is no end to the mathematics that can be observed as the light beams do their dance on the string model. (Many people have suggested that a musical accompaniment should be provided as an added dimension as the curves dance on the strings.)

"Dancing Curves" will prove to be an exciting visual experience, limited only by the imagination of the lecturer and the observer.